Table of Contents

RMF ISSO:
FOUNDATIONS (GUIDE)

NIST 800 RISK MANAGEMENT FRAMEWORK FOR CYBERSECURITY PROFESSIONALS

BRUCE BROWN

RMF ISSO:
Foundations (Guide)
Book 1 (v1)

NIST 800 Risk Management Framework For Cybersecurity Professionals

By Bruce Brown, CISSP, ISC2 CAP

Convocourses Contacts:

Video Courses:

convocourses.com

youtube.com/convocourses

Services:

contact@convocourses.com

Getting Started

This book is a practical guide to learn how to be an **Information System Security Officer (ISSO)** doing the **risk management framework (RMF)** process following *NIST SP 800-37, Risk Management Framework for Information Systems and Organizations: A System Life Cycle Approach for Security and Privacy*. This guide is foundational and basic but may be used by information technology and cyber security professionals who are about to take an ISSO role and need to know about RMF. While it can be used by risk management experts who need a refresher, I've written it so that a person who has little or no IT or cybersecurity compliance experience will know what needs to be done.

Who is this book for?

- People who want to learn about the ISSO role

- People who want to learn about security compliance for NIST 800-37 (RMF)

- People who want to get a job as an ISSO

- Computer hobbyists trying to expand their knowledge of security compliance

- IT professionals

- Cybersecurity professionals

Learning NIST Risk Management Framework can help you understand other security frameworks. For example, there is an international standard that is similar to the NIST Risk Management Framework called ISO/IEC 27000, Information Security Management Systems (ISMS).

If you are looking for an information system security officer (ISSO) or cyber security compliance position, you are in luck! Because at the time of this writing there is a shortage of people who can do risk management framework or have the skills to do it. This practical guide gives the essentials of what I have learned while doing risk management framework as an ISSO.

What is an ISSO?

Let's start with the basics. What is an ISSO?

The Department of Homeland Security (DHS) Information System Security Officer Guide describes an ISSO in the following way:

Information System Security Officer (ISSO) serves as the principal advisor to the Information System Owner (SO), Business Process Owner, and the Chief Information Security Officer (CISO) / Information System Security Manager (ISSM) on all matters, technical and otherwise, involving the security of an information system. ISSOs are responsible for ensuring the implementation and maintenance of security controls in accordance with the Security Plan (SP) and Department of Homeland Security (DHS) policies. -dhs.gov

Here is how the US Department of the Navy describes the ISSO role:

The ISSO is formally appointed in writing by the program manager of a specific branch, division, or department, as appropriate, based on the structure and needs of the specific command or activity. The Information System Security Manager (ISSM) provides input to the program manager regarding the appointment decision. If requested, the ISSM may provide technical help in the development of appointment memos or letters. The ISSO appointment letter briefly summarizes the duties and responsibilities of the ISSO. Depending on the Command structure, over one ISSO may be appointed. Commands having complex ISs may need more ISSOs to perform day-to-day activities and to respond to security problems and IS user needs. - SECNAV M-5510.36

In my experience, the ISSO's job is to help manage the overall risk to the organization by managing the security of the information systems and environment.

ISSO Role in Each Organization

I have done ISSO work for many organizations, including NASA, the DoD, and the private sector. While there are differences in the tasks and inter-agency processes, there are some things that don't change:

● Coordination with stakeholders (i.e., system administrators, management, engineers)

● Document the security features of systems

- Identify system assets and components

- Assist in tracking new threats and vulnerabilities

In every ISSO position I have held, I have had to coordinate with the system administrator/engineers (the person implementing security controls and designing the system), the managers (supervisors and account managers, and the client (the system owners). It usually includes creating or managing security documentation, such as system security plans. An essential aspect of the job requires knowing what the system assets and components are and a general understanding of the purpose of the system.

The ISSO coordinates with all the system stakeholders to help develop an acceptable interpretation of the computer security implemented (security controls), the system security policies, and regulations. This interpretation gives valuable guidance to help manage the risk level of the system. The goal is to limit and manage the risks to the organization. It is not realistic to eliminate 100% of the risk associated with any system, but you can **manage** it.

What is RMF?

RMF stands for risk management framework. From the NIST.gov's website, RMF is:

> The selection and specification of security controls for a system is accomplished as part of an organization-wide information security program that involves the management of organizational risk—-that is, the risk to the organization or to individuals associated with the operation of a system. The management of organizational risk is a key element in the organization's information security program and provides an effective framework for selecting the appropriate security controls for a system—-the security controls necessary to protect individuals and the operations and assets of the organization.

In layperson's terms, the RMF is a process for *minimizing the potential negative impacts* of a system on an organization by ensuring security controls are implemented properly.

That's it. Notice the focus is on the **risk to the organization**. You protect systems and individuals in the organization, limiting the risks of something bad happening to the organization or ensuring that the organization can respond when something bad happens. *Something bad = "negative impact" = (hacks, breaches, malware, natural disasters, user errors, theft, leaked data, power outages, mechanical errors, etc.).*

There are many kinds of risk management frameworks across multiple industries. Countries have their standards; there are international standards and standards in different industries. National Institute of Standards and Technology (NIST) created a risk management framework that the United States uses. I detailed the entire process in their document, *NIST Special Publication (SP) 800-37, Risk Management Framework for Information Systems and Organizations: A System Life Cycle Approach for Security and Privacy*. It is called NIST RMF for short.

Minimizing Negative impacts

The NIST RMF has a seven-step process for minimizing the negative impacts on an asset. It is important to note that this process is done for each group of systems and assets within a system boundary (i.e., a local area network, a weapons system, a program management system, an enterprise, or any local and physical boundary defined and controlled by a head of the agency.). Here are the seven steps:

1. Prepare

2. Categorize

3. Select

4. Implement

5. Assess

6. Authorize

7. Monitor

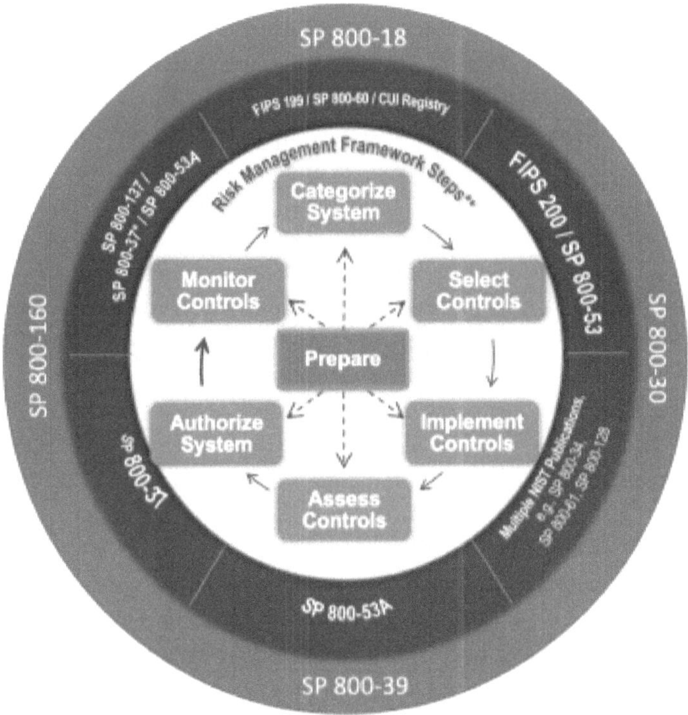

Prepare. The Prepare step is the foundation of all other steps because data is gathered at all enterprise levels to start the RMF process for a specific system. All the stakeholders for the system are listed and contacted in preparation for the risk management framework process. The goal is to prepare the organization to manage its security and privacy risks using the Risk Management Framework.

Categorize. In this step, the organization will categorize the security and impact level of the system and the information processed, stored, and transmitted based on impact analysis. (Reference: FIPS 199, NIST SP 800-60, NIST SP 800-30, NIST SP 800-18)

Select. The organization will select an initial set of baseline security controls for the system. This depends on the security categorization. During the selection step, the organization also tailors and supplements the security controls as needed based on the organization's assessment of risk and local conditions. (Reference: FIPS 200, NIST SP 800-53)

Implement. The organization will implement the security controls and document how the controls are deployed within the system and environment of operation. (Reference: Security Technical Implementation Guides, vendor documents/sites)

Asset. The organization will ensure the systems controls are assessed. They assess the security controls using procedures to determine if the controls are implemented correctly, operating as intended, and producing the desired outcome concerning meeting the security requirements for the system. (Reference: NIST SP 800-53A)

Authorize. After the system security is assessed, it needs to be authorized. Authorization is based upon determining the risk to the organization's operations, assets, individuals, and the nation. The head of the organization decides if the risk is acceptable. (Reference: NIST SP 800-37)

Monitor. Monitor and assess selected security controls in the system on an ongoing basis this includes assessing security control effectiveness, documenting changes to the system or environment of operation, conducting security impact analysis of the associated changes, and reporting the system's security state to appropriate organizational officials. (Reference: NIST SP 800-37, NIST SP 800-137, NIST SP 800-53A).

Source: csrc.nist.gov

RMF ISSO Foundation

The NIST RMF process is based on regulations, laws, and engineering standards. It incorporates concepts from the NIST *Framework for Improving Critical Infrastructure Cybersecurity* that complements the currently established risk management processes mandated by the Office of Management and Budget and the Federal Information Security Modernization Act (FISMA). In July 2016, the Office of Management and Budget (OMB) revised Circular A-130 to require federal agencies to apply the RMF within their privacy programs.

All the processes and laws of risk management come down to protecting the security objectives of the information on the system. The security objectives are:

- Confidentiality – restricting who has access to the information.
- Integrity – ensuring the original information is not manipulated or corrupted.
- Availability – making the data accessible to those who need it.

As an information system security officer, focusing on these three things will help you understand why certain controls are in place and how to interpret them when you read the NIST guidance.

The NIST SP 800-37 RMF uses the security controls categorized in NIST SP 800-53. There are hundreds of security and privacy controls, and they are ever adjusting, but here is what the families look like:

AC - ACCESS CONTROL[1]
AT - AWARENESS AND TRAINING[2]
AU - AUDIT AND ACCOUNTABILITY[3]

---

1. https://csrc.nist.gov/Projects/risk-management/sp800-53-controls/

   release-search#_6666cd76f96956469e7be39d750cc7d9_controls_6666cd76f96956469e7be39d7

   50cc7d9__d1457b72c3fb323a2671125aef3eab5d_version_43ec3e5dee6e706af7766fffea512721_

   5.1_6cff047854f19ac2aa52aac51bf3af4a_family_43ec3e5dee6e706af7766fffea512721_AC

2. https://csrc.nist.gov/Projects/risk-management/sp800-53-controls/

   release-search#_6666cd76f96956469e7be39d750cc7d9_controls_6666cd76f96956469e7be39d7

   50cc7d9__d1457b72c3fb323a2671125aef3eab5d_version_43ec3e5dee6e706af7766fffea512721_

   5.1_6cff047854f19ac2aa52aac51bf3af4a_family_43ec3e5dee6e706af7766fffea512721_AT

<u>CA - ASSESSMENT, AUTHORIZATION, AND MONITORING[4]</u>
<u>CM - CONFIGURATION MANAGEMENT[5]</u>
<u>CP - CONTINGENCY PLANNING[6]</u>
<u>IA - IDENTIFICATION AND AUTHENTICATION[7]</u>
<u>IR - INCIDENT RESPONSE[8]</u>
<u>MA - MAINTENANCE[9]</u>
<u>MP - MEDIA PROTECTION[10]</u>

---

3. https://csrc.nist.gov/Projects/risk-management/sp800-53-controls/
release-search#_6666cd76f96956469e7be39d750cc7d9_controls_6666cd76f96956469e7be39d7
50cc7d9__d1457b72c3fb323a2671125aef3eab5d_version_43ec3e5dee6e706af7766fffea512721_
5.1_6cff047854f19ac2aa52aac51bf3af4a_family_43ec3e5dee6e706af7766fffea512721_AU

4. https://csrc.nist.gov/Projects/risk-management/sp800-53-controls/
release-search#_6666cd76f96956469e7be39d750cc7d9_controls_6666cd76f96956469e7be39d7
50cc7d9__d1457b72c3fb323a2671125aef3eab5d_version_43ec3e5dee6e706af7766fffea512721_
5.1_6cff047854f19ac2aa52aac51bf3af4a_family_43ec3e5dee6e706af7766fffea512721_CA

5. https://csrc.nist.gov/Projects/risk-management/sp800-53-controls/
release-search#_6666cd76f96956469e7be39d750cc7d9_controls_6666cd76f96956469e7be39d7
50cc7d9__d1457b72c3fb323a2671125aef3eab5d_version_43ec3e5dee6e706af7766fffea512721_
5.1_6cff047854f19ac2aa52aac51bf3af4a_family_43ec3e5dee6e706af7766fffea512721_CM

6. https://csrc.nist.gov/Projects/risk-management/sp800-53-controls/
release-search#_6666cd76f96956469e7be39d750cc7d9_controls_6666cd76f96956469e7be39d7
50cc7d9__d1457b72c3fb323a2671125aef3eab5d_version_43ec3e5dee6e706af7766fffea512721_
5.1_6cff047854f19ac2aa52aac51bf3af4a_family_43ec3e5dee6e706af7766fffea512721_CP

7. https://csrc.nist.gov/Projects/risk-management/sp800-53-controls/
release-search#_6666cd76f96956469e7be39d750cc7d9_controls_6666cd76f96956469e7be39d7
50cc7d9__d1457b72c3fb323a2671125aef3eab5d_version_43ec3e5dee6e706af7766fffea512721_
5.1_6cff047854f19ac2aa52aac51bf3af4a_family_43ec3e5dee6e706af7766fffea512721_IA

8. https://csrc.nist.gov/Projects/risk-management/sp800-53-controls/
release-search#_6666cd76f96956469e7be39d750cc7d9_controls_6666cd76f96956469e7be39d7
50cc7d9__d1457b72c3fb323a2671125aef3eab5d_version_43ec3e5dee6e706af7766fffea512721_
5.1_6cff047854f19ac2aa52aac51bf3af4a_family_43ec3e5dee6e706af7766fffea512721_IR

<u>PE - PHYSICAL AND ENVIRONMENTAL PROTECTION[11]</u>
<u>PL - PLANNING[12]</u>
<u>PM - PROGRAM MANAGEMENT[13]</u>
<u>PS - PERSONNEL SECURITY[14]</u>
<u>PT - PERSONALLY IDENTIFIABLE INFORMATION PROCESSING AND TRANSPARENCY[15]</u>
<u>RA - RISK ASSESSMENT[16]</u>

---

9. https://csrc.nist.gov/Projects/risk-management/sp800-53-controls/
release-search#_6666cd76f96956469e7be39d750cc7d9_controls_6666cd76f96956469e7be39d7
50cc7d9__d1457b72c3fb323a2671125aef3eab5d_version_43ec3e5dee6e706af7766fffea512721_
5.1_6cff047854f19ac2aa52aac51bf3af4a_family_43ec3e5dee6e706af7766fffea512721_MA

10. https://csrc.nist.gov/Projects/risk-management/sp800-53-controls/
release-search#_6666cd76f96956469e7be39d750cc7d9_controls_6666cd76f96956469e7be39d7
50cc7d9__d1457b72c3fb323a2671125aef3eab5d_version_43ec3e5dee6e706af7766fffea512721_
5.1_6cff047854f19ac2aa52aac51bf3af4a_family_43ec3e5dee6e706af7766fffea512721_MP

11. https://csrc.nist.gov/Projects/risk-management/sp800-53-controls/
release-search#_6666cd76f96956469e7be39d750cc7d9_controls_6666cd76f96956469e7be39d7
50cc7d9__d1457b72c3fb323a2671125aef3eab5d_version_43ec3e5dee6e706af7766fffea512721_
5.1_6cff047854f19ac2aa52aac51bf3af4a_family_43ec3e5dee6e706af7766fffea512721_PE

12. https://csrc.nist.gov/Projects/risk-management/sp800-53-controls/
release-search#_6666cd76f96956469e7be39d750cc7d9_controls_6666cd76f96956469e7be39d7
50cc7d9__d1457b72c3fb323a2671125aef3eab5d_version_43ec3e5dee6e706af7766fffea512721_
5.1_6cff047854f19ac2aa52aac51bf3af4a_family_43ec3e5dee6e706af7766fffea512721_PL

13. https://csrc.nist.gov/Projects/risk-management/sp800-53-controls/
release-search#_6666cd76f96956469e7be39d750cc7d9_controls_6666cd76f96956469e7be39d7
50cc7d9__d1457b72c3fb323a2671125aef3eab5d_version_43ec3e5dee6e706af7766fffea512721_
5.1_6cff047854f19ac2aa52aac51bf3af4a_family_43ec3e5dee6e706af7766fffea512721_PM

14. https://csrc.nist.gov/Projects/risk-management/sp800-53-controls/
release-search#_6666cd76f96956469e7be39d750cc7d9_controls_6666cd76f96956469e7be39d7
50cc7d9__d1457b72c3fb323a2671125aef3eab5d_version_43ec3e5dee6e706af7766fffea512721_
5.1_6cff047854f19ac2aa52aac51bf3af4a_family_43ec3e5dee6e706af7766fffea512721_PS

<u>SA - SYSTEM AND SERVICES ACQUISITION[17]</u>
<u>SC - SYSTEM AND COMMUNICATIONS PROTECTION[18]</u>
<u>SI - SYSTEM AND INFORMATION INTEGRITY[19]</u>
<u>SR - SUPPLY CHAIN RISK MANAGEMENT[20]</u>

Source: csrc.nist.gov

---

15. https://csrc.nist.gov/Projects/risk-management/sp800-53-controls/
release-search#_6666cd76f96956469e7be39d750cc7d9_controls_6666cd76f96956469e7be39d7
50cc7d9__d1457b72c3fb323a2671125aef3eab5d_version_43ec3e5dee6e706af7766fffea512721_
5.1_6cff047854f19ac2aa52aac51bf3af4a_family_43ec3e5dee6e706af7766fffea512721_PT

16. https://csrc.nist.gov/Projects/risk-management/sp800-53-controls/
release-search#_6666cd76f96956469e7be39d750cc7d9_controls_6666cd76f96956469e7be39d7
50cc7d9__d1457b72c3fb323a2671125aef3eab5d_version_43ec3e5dee6e706af7766fffea512721_
5.1_6cff047854f19ac2aa52aac51bf3af4a_family_43ec3e5dee6e706af7766fffea512721_RA

17. https://csrc.nist.gov/Projects/risk-management/sp800-53-controls/
release-search#_6666cd76f96956469e7be39d750cc7d9_controls_6666cd76f96956469e7be39d7
50cc7d9__d1457b72c3fb323a2671125aef3eab5d_version_43ec3e5dee6e706af7766fffea512721_
5.1_6cff047854f19ac2aa52aac51bf3af4a_family_43ec3e5dee6e706af7766fffea512721_SA

18. https://csrc.nist.gov/Projects/risk-management/sp800-53-controls/
release-search#_6666cd76f96956469e7be39d750cc7d9_controls_6666cd76f96956469e7be39d7
50cc7d9__d1457b72c3fb323a2671125aef3eab5d_version_43ec3e5dee6e706af7766fffea512721_
5.1_6cff047854f19ac2aa52aac51bf3af4a_family_43ec3e5dee6e706af7766fffea512721_SC

19. https://csrc.nist.gov/Projects/risk-management/sp800-53-controls/
release-search#_6666cd76f96956469e7be39d750cc7d9_controls_6666cd76f96956469e7be39d7
50cc7d9__d1457b72c3fb323a2671125aef3eab5d_version_43ec3e5dee6e706af7766fffea512721_
5.1_6cff047854f19ac2aa52aac51bf3af4a_family_43ec3e5dee6e706af7766fffea512721_SI

20. https://csrc.nist.gov/Projects/risk-management/sp800-53-controls/
release-search#_6666cd76f96956469e7be39d750cc7d9_controls_6666cd76f96956469e7be39d7
50cc7d9__d1457b72c3fb323a2671125aef3eab5d_version_43ec3e5dee6e706af7766fffea512721_
5.1_6cff047854f19ac2aa52aac51bf3af4a_family_43ec3e5dee6e706af7766fffea512721_SR

Prepare

"Prepare" was added to revision 2 of the NIST 800-37. It is now the first step of the risk management framework. The purpose is to prepare the organization to go through the process of applying security and privacy to a system.

In the past, before "prepare" was added, most information system security officers and organizations "prepared" in some way before starting. When there is no preparation, many people within the organization are caught off guard or don't have the resources even to start the RMF process. The Prepare step gathers all the stakeholders (those with some responsibility for the system you are working on) and gets everyone on the same page. The stakeholders are told what is going on, who is involved, who we need, where it will be when it is complete, and why we even have this system.

The System Security Plan: As an ISSO, you will start drafting the system security plan (SSP) during the preparation step.

Objectives of preparing a system for the RMF process:

- Assign roles for the risk management process

- Have a risk management process

- Identify what missions/businesses the system supports

- Identify all stakeholders

- Identify the asset

- Identify the personally identifiable data process (if applicable)

- Conduct a risk assessment or update existing ones

- Identify and prioritize security and privacy requirements of the system

- Determine where the system will be within the enterprise architecture

- Define organizational security and privacy architecture

- Identify and publish common controls that are available for inheritance

- Prioritize similar systems among other systems in the organization

- Tailored security and privacy control baselines are established and made available for enterprise-wide use.

- Set up an organizational plan for monitoring security and privacy controls that are in place

Assign Roles Associated & System Stakeholders

The NIST 800-37 assigns this task to the "Head of Agency" or the "Chief Information Officer" or "Senior Agency Official for Privacy." Your organization will have some of these roles already assigned. You may find these roles in the security policy. The security policy is sometimes called a system security policy or an information system security plan, but many other names exist. Essentially it is a document that breaks down all of the organization's security/privacy practices.

You can also find the assigned roles in an organizational chart. If there is no documentation of the assigned roles, contact your organization to determine the roles.

Here are some of the roles that many organizations include when going through the RMF process for an information system. Keep in mind that the names of the roles can be different at each organization, and some of the roles will not be necessary for some systems. But in general, these are the kinds of roles you will see:

Authorizing Official (AO) - Responsible for the entire system. Gives final approval for systems and takes formal responsibility for risks in writing. Either above the CIO, the CIO or appointed by the CIO

(aka - DAA, Designated Approval Authority, Commander, CIO, AO). **From the ISSO's perspective:** The AO sometimes has an office or representative that works with the ISSO. These are arguably the most important roles that will need to be a part of the process. The AO approves, disapproves, or takes the risk of the system going through the RMF process. You may not ever see them personally, but their orders and direction are why you are analyzing the security of the system and creating documents.

Common Control Provider - Individual or group responsible for implementing, assessing, and monitoring common controls. Common controls are security controls multiple resources in the organization uses in the same way. Physical security controls, like 24/7 security units, are considered common controls. Common Controls can be inherited from the organization. For example, a laptop inherits the organization's physical security once inside the building. Another inherited control might be that all systems in the facility are plugged into an uninterrupted power supply. **From the ISSO's perspective:** The Common Control Provider is usually explained in the organization's policies. The security control policy should explain physical security, personnel security, access controls, and other things that apply to ALL people and systems in the organization. If there is NO security control policy, the ISSO and ISSM may need to get with the information system owner and customer and create one. That document (or a document like it) will cover common controls.

Control Assessor - Individual or group that conducts a comprehensive assessment of the security controls. Need some level of independence from the authorizing official. Sometimes it is a unit working for the AO (risk management team, assessment team, risk assessment team, auditors, audit team). **From the ISSO's perspective:** The Control Assessors are scary people because they report problems directly to the CIO/AO. But if you do what you're supposed to do, there is less stress. They can be your best asset. If you ask them what they need from you before they even get there, they are usually willing to give you a checklist and forms to fill out, and sometimes they will even glance at your system before they start officially and tell you how to do the fix.

Information Owner Or Steward - This is an organizational official with statutory, management, or operational authority for specified information. They are responsible for ensuring that information follows policy and procedures when generated, collected, processed, stored, and disposed of (aka information system officer, ISO, data owner, system owner). This is very similar to a **Mission/Business Owner** whose focus is a specific line of business (contract supporting a system). **From the ISSO's perspective:** The information owner will need to be informed on every process level. They are like representatives of the system. They speak to the office of the AO, the users of the system, and you (the ISSO). They are one of the most influential people you need to talk to because they know what AO, Assessors, and users want to see. These people are not always called Information Owners. But you will know them because they are not entirely in charge but in the middle of everything involving that system. If you have one of these people in your organization, keep them in the loop.

System Security / Privacy Officers - An organization should have a system/privacy officer who ensures each system fits with the organization's operation security/privacy posture. **From the ISSO's perspective:** The ISSO might work with other system security/privacy officers to perform privacy impact assessments or privacy threshold analyses.

Senior Accountable Official for Risk Management - It is a fancy title for the person (or group) in your organization that leads the direction of the risk management process. In most cases, this is appointed by the head of the organization and under the office of the CIO. They lead the Risk Management (function). It is called a risk management "function" because it is not necessarily a single group or person, but it is something that must be done regardless of group, title, or role.

It is a process. **From the ISSO's perspective:** This could play a significant role in that all processes and documentation might go through this person, their representatives, or their office. They will look through your documentation, evidence, and process to determine if you are in line with the rest of the organization before sending the documents forward to the AO. An ISSO wants to make sure they are entirely in line with the office that is in charge of the risk management process.

Senior Agency Information Security Officer - They are an organizational official responsible for carrying out the chief information officer security responsibilities under FISMA and serving as the primary liaison for the chief information officer to the organization's authorizing officials, system owners, common control providers, and system security officers. They usually have several hats, including but not limited to: risk executive function, head of common control providers, and system owners. **From the ISSO's perspective:** This position is defined by what they do. It will usually be high-level security that does *everything* security related for the CIO. Sometimes it is the CIO or part of the office of the CIO. They are sometimes the main people you talk to for privacy, common controls, security control assessments, risk executive functions... and anything else you can think of related to security at the executive level.

System Administrator - The System Administrator is an individual, group, or organization responsible for setting up and maintaining a system or specific components of a system. System administrator responsibilities include: installing, configuring, and updating hardware and software; establishing and managing user accounts. This is the person who touches the system. **From the ISSO's perspective:**

Nothing on the system will get done without the system administrator. You need to be in contact with this person or organization from day one, even if it is just to let them know what is going to happen. In some cases, the ISSO is also doing some system administrator work.

System Owner - A system owner is an organizational official responsible for the procurement, development, integration, modification, operation, maintenance, and disposal of an organizational system. The system owner is responsible for addressing the operational interests of the user community (i.e., users who require access to the system to satisfy mission, business, or operational requirements) and for ensuring compliance with security requirements. **From the ISSO's perspective:** The information system owner, information owner, and system owner are interchangeable in some organizations. The system owner makes sure that the user experience is not interrupted by the security. They also want to ensure that there is adequate security in place. The system owner may conduct an operational test to ensure that the system works appropriately after all security controls are applied.

See the end of this guide for more roles.

System Stakeholders

During the Prepare step, the system will need a list of stakeholders. This will be a point of a contact list that covers some or all the roles identified for the system and organization.

Example of a stakeholder list:

> Capt. James T. Kirk - Authorization Officers - JamesKirk@fleet.mz.gov
>
> Commander Burnham - AO representative-Burnham@fleet.mz.gov
>
> Maggie Van Wrinkle T - System Owner - MVTWringle@fleet.mz.gov
>
> Dr. Sean D. Cochran - System Admin - Sean.Cochran@fleet.mz.gov

You want a point of contact list with the most relevant stakeholders. With small organizations, many of these stakeholders may be the same people. The stakeholder's list is just for you. It will come in handy. You may need it to populate official documents.

References: *NIST Special Publications800-39[1] (Organization Level),800-64[2],800-160[3] (Stakeholder Needs and Requirements Definition and Portfolio Management Processes),800-161[4];Cybersecurity[5]Framework[6].*

---

1. http://nvlpubs.nist.gov/nistpubs/Legacy/SP/nistspecialpublication800-39.pdf

2. http://nvlpubs.nist.gov/nistpubs/Legacy/SP/nistspecialpublication800-64r2.pdf

3. http://nvlpubs.nist.gov/nistpubs/SpecialPublications/NIST.SP.800-160.pdf

4. http://doi.org/10.6028/NIST.SP.800-161

5. https://www.nist.gov/sites/default/files/documents/cyberframework/cybersecurity-framework-021214.pdf

6. https://www.nist.gov/sites/default/files/documents/cyberframework/cybersecurity-framework-021214.pdf

Establish a Risk Management Strategy for the Organization

Part of the "prepare" step is making sure the organization has a risk management strategy. This is important for the entire RMF process because it shows that the organization has agreed to follow the RMF process in writing.

The organization needs to have documentation that details a risk management strategy. It should explain the who, what, when, and where of risk management for the organization. This is **not** something the ISSO does in isolation. The NIST SP 800-37 states that this is the responsibility of the **head of the agency** (C-Level executives like CEO, CIO, General, commander, upper-level management given the responsibility, etc.). They usually task someone to do it, but the head of the agency must approve and sign off on it.

The risk management strategy may be detailed in the organization's security plan or security policy. The document should describe how the organization implements the risk management process. It should explain risk tolerance, the level or degree of risk, or uncertainty acceptable to an organization. For example, suppose you have a new system installed in the organization. In that case, you need to have a plan to determine if it will introduce new vulnerabilities to the organization. And you will need to have a process that can help determine if the new system has a low, moderate, or high-risk impact on the organization. A risk management strategy will explain how to deal with new risks, measure them, and how the organization deals with each level of risk based on their impact on the organization.

The organization's risk management strategy (sometimes tied to the "security policy") should include the RMF process that addresses how to continually assess systems (new and old), evaluate them, and measure their risk level. This is something that senior leadership and executives throughout the organization need to push. While the ISSO can create one, the authority behind this document comes from above. If you are an ISSO and the organization does not have one, you should let them know that it is a federal regulation to have a risk management strategy (*sources: Executive[1] Order[2] 13800[3]; Federal[4] Information[5] Security[6] Modernization[7] Act[8]; Homeland[9] Security[10] Presidential[11] Directive[12] 7[13]; OMB[14] Circular[15] A[16]-130[17]*). The risk management strategy will align with the NIST SP 800-37.

References: NIST Special Publications 800-30[18], 800-39[19] (Organization Level), 800-160[20] (Risk Management, Quality Assurance, Quality Management, Decision Management, Project Assessment, and

1.   *https://csrc.nist.gov/Topics/Laws-and-Regulations/executive-documents/EO-13800*

2.   *https://csrc.nist.gov/Topics/Laws-and-Regulations/executive-documents/EO-13800*

3.   *https://csrc.nist.gov/Topics/Laws-and-Regulations/executive-documents/EO-13800*

4.   *https://csrc.nist.gov/Topics/Laws-and-Regulations/laws/FISMA*

5.   *https://csrc.nist.gov/Topics/Laws-and-Regulations/laws/FISMA*

6.   *https://csrc.nist.gov/Topics/Laws-and-Regulations/laws/FISMA*

7.   *https://csrc.nist.gov/Topics/Laws-and-Regulations/laws/FISMA*

8.   *https://csrc.nist.gov/Topics/Laws-and-Regulations/laws/FISMA*

9.   *https://csrc.nist.gov/Topics/Laws-and-Regulations/executive-documents/HSPD-7*

10.   *https://csrc.nist.gov/Topics/Laws-and-Regulations/executive-documents/HSPD-7*

11.   *https://csrc.nist.gov/Topics/Laws-and-Regulations/executive-documents/HSPD-7*

12.   *https://csrc.nist.gov/Topics/Laws-and-Regulations/executive-documents/HSPD-7*

13.   *https://csrc.nist.gov/Topics/Laws-and-Regulations/executive-documents/HSPD-7*

14.   *https://csrc.nist.gov/Topics/Laws-and-Regulations/executive-documents/OMB-A-130*

15.   *https://csrc.nist.gov/Topics/Laws-and-Regulations/executive-documents/OMB-A-130*

16.   *https://csrc.nist.gov/Topics/Laws-and-Regulations/executive-documents/OMB-A-130*

17.   *https://csrc.nist.gov/Topics/Laws-and-Regulations/executive-documents/OMB-A-130*

18.   *http://doi.org/10.6028/NIST.SP.800-30r1*

19.   *http://nvlpubs.nist.gov/nistpubs/Legacy/SP/nistspecialpublication800-39.pdf*

20.   *http://nvlpubs.nist.gov/nistpubs/SpecialPublications/NIST.SP.800-160.pdf*

Control Processes), 800-161[21]; NIST Interagency Report 8062[22]; Cybersecurity[23] Framework[24].

21.   *http://doi.org/10.6028/NIST.SP.800-161*

22.   *http://nvlpubs.nist.gov/nistpubs/ir/2017/NIST.IR.8062.pdf*

23.   *https://www.nist.gov/sites/default/files/documents/cyberframework/cybersecurity-framework-021214.pdf*

24.   *https://www.nist.gov/sites/default/files/documents/cyberframework/cybersecurity-framework-021214.pdf*

Conduct an Initial Risk Assessment of Assets

During the preparation process, the ISSO should consider looking at the risks associated with the system. Take a look at any old vulnerability scan, risk assessments, compliance scans, or any other assessment done on the system. This will allow you to see where the system is from a security perspective. If there has not been any recent assessment done, you should conduct a self-assessment of the system using any available resources. Check out logs, do a vulnerability scan, do a fresh set of assessments on the policies and vulnerabilities, and talk to the engineers about the system's current status. These actions allow you to get prepared to categorize the system's security.

Now let's say there is *no system*! What do you do in the preparation stage for a system that does not exist yet? Look at the designs, diagrams, and architectures of the system.

Some organizations have risk assessment teams that can conduct an initial risk assessment. Assessment of risk includes identifying threat sources and threat events affecting stakeholder assets, how the assets are vulnerable to the threats, the likelihood that a threat source/event will exploit an asset vulnerability, and the impact (consequence) of loss of the assets.

Throughout the system development life cycle, risk assessments are conducted and support various RMF steps and tasks. Risk assessment results can guide and inform potential courses of action for risk responses. Organizations determine the form of the risk assessment conducted (including the scope, depth, and formality of such assessments) and the method of reporting results.

The risk assessment report will give the organization an idea of where the system is regarding the level of risk. It is good to run scans and review the documentation and existing security controls in the preparation phase.

References: *FIPS Publications 199[1], 200[2]; NIST Special Publications 800-30,[3] 800-39[4] (Organization Level), 800-59[5], 800-60[6], 800-64[7], 800-160[8] (Stakeholder Needs and Requirements Definition and Risk Management Processes), 800-161[9] (Assess); NIST Interagency Reports 8062[10], 8179[11];Cybersecurity[12]Framework[13] (Core Identify Function); CNSS Instruction 1253.*

1.   *http://nvlpubs.nist.gov/nistpubs/FIPS/NIST.FIPS.199.pdf*

2.   *http://nvlpubs.nist.gov/nistpubs/FIPS/NIST.FIPS.200.pdf*

3.   *http://nvlpubs.nist.gov/nistpubs/Legacy/SP/nistspecialpublication800-30r1.pdf*

4.   *http://nvlpubs.nist.gov/nistpubs/Legacy/SP/nistspecialpublication800-39.pdf*

5.   *http://nvlpubs.nist.gov/nistpubs/Legacy/SP/nistspecialpublication800-59.pdf*

6.   *http://nvlpubs.nist.gov/nistpubs/Legacy/SP/nistspecialpublication800-60v1r1.pdf*

7.   *http://nvlpubs.nist.gov/nistpubs/Legacy/SP/nistspecialpublication800-64r2.pdf*

8.   *http://nvlpubs.nist.gov/nistpubs/SpecialPublications/NIST.SP.800-160.pdf*

9.   *http://doi.org/10.6028/NIST.SP.800-161*

10.   *http://nvlpubs.nist.gov/nistpubs/ir/2017/NIST.IR.8062.pdf*

11.   *http://csrc.nist.gov/publications/drafts/nistir-8179/nistir-8179-draft.pdf*

12.   *https://www.nist.gov/sites/default/files/documents/cyberframework/cybersecurity-framework-021214.pdf*

13.   *https://www.nist.gov/sites/default/files/documents/cyberframework/cybersecurity-framework-021214.pdf*

Prepare for the System to Inherit Common Controls

During the preparation phase, figure out what common controls apply to the system and document them.

Common controls are established by the organization by a senior information security officer. As described above, common controls are security and privacy controls that apply to more than one system within an organization. They are controls that one or more organizational system can inherit. Such controls can include physical and environmental protection controls, personnel security controls, or complaint management controls for receiving privacy-related inquiries from the public. Organizations identify and select the set of common controls and assign those controls to people or groups designated as common control providers. Common controls may differ based on various factors, such as hosting location, system architecture, and organizational structure. The list of common controls should take these factors into account. Common controls can also be identified at different levels of the organization—for example, corporate or agency level, bureau or subcomponent level, or individual department level. Organizations may establish one or more lists of common controls that organizational systems can inherit.

From the ISSO's perspective, if there are NO common controls published, inform the organization of the need to put them in a document such as security policy. Some organizations have the entire RMF process tracked in a database; these databases will identify the common controls. If there are no common controls identified, let the organization know that they need to identify the common controls and use NIST SP 800-53 as a reference.

References: *NIST Special Publication 800-53[1].*

---

1. *http://nvlpubs.nist.gov/nistpubs/SpecialPublications/NIST.SP.800-53r4.pdf*

Prepare Continuous Monitoring

During the preparation step, the ISSO should check the organization's continuous monitoring strategy and document how it impacts the system. If this is already documented, see if the assessments' frequency is on the system. For example, take note of the following privacy impact assessment or the next vulnerability scan on the system.

It's the responsibility of the organization to monitor the security and privacy controls. Continuous monitoring within the organization is initiated by one or more of the following roles: Senior Agency Information Security Officer; Senior Agency Official for Privacy; Senior Accountable Official for Risk Management/Risk Executive (Function); Chief Information Officer.

The ISSO is part of the continuous monitoring strategy. This is nothing more than "how often does the organization check the security controls on System A?" and "How will monitoring be done?" The monitoring strategy also identifies the minimum frequency of monitoring for implemented controls across the organization and defines the organizational control assessment approach. The organizational monitoring strategy may also define how changes to systems are monitored, how security and privacy impact analyses are conducted, and security and privacy status reporting requirements, including recipients of the status, reports.

References: NIST Special Publications 800-30[1], 800-39[2] (Organization, Mission/Business Process, System Levels), 800-53[3], 800-53[4]A[5], 800-161[6], 800-137[7]; NIST Interagency Report

1.   http://nvlpubs.nist.gov/nistpubs/Legacy/SP/nistspecialpublication800-30r1.pdf

2.   http://nvlpubs.nist.gov/nistpubs/Legacy/SP/nistspecialpublication800-39.pdf

3.   http://nvlpubs.nist.gov/nistpubs/SpecialPublications/NIST.SP.800-53r4.pdf

4.   http://nvlpubs.nist.gov/nistpubs/SpecialPublications/NIST.SP.800-53Ar4.pdf

5.   http://nvlpubs.nist.gov/nistpubs/SpecialPublications/NIST.SP.800-53Ar4.pdf

6.   http://doi.org/10.6028/NIST.SP.800-161

8062[8];Cybersecurity[9]Framework[10] (Core Detect Function); CNSS Instruction 1253.

---

7.   *http://nvlpubs.nist.gov/nistpubs/Legacy/SP/nistspecialpublication800-137.pdf*

8.   *http://nvlpubs.nist.gov/nistpubs/ir/2017/NIST.IR.8062.pdf*

9.   *https://www.nist.gov/sites/default/files/documents/cyberframework/cybersecurity-framework-021214.pdf*

10.   *https://www.nist.gov/sites/default/files/documents/cyberframework/cybersecurity-framework-021214.pdf*

Identify the Missions, Business Functions

In the preparation step of the RMF, the organization should identify the mission/business functions the system is intended to support. This is USUALLY understood before work begins. Depending on the situation, technical system documentation will explain the system itself and what mission and business it supports.

For new systems, it is imperative to know what it supports. The more access the ISSO has to the concept of operations documents, the better.

References: NIST Special Publications800-39[1] (Organization and Mission/Business Process Levels),800-64[2],800-160[3] (Business or Mission Analysis, Portfolio Management, and Project Planning Processes); NIST Interagency Report8179[4] (Criticality Analysis Process B);Cybersecurity[5]Framework[6] (Core Identify Function).

---

1.   http://nvlpubs.nist.gov/nistpubs/Legacy/SP/nistspecialpublication800-39.pdf

2.   http://nvlpubs.nist.gov/nistpubs/Legacy/SP/nistspecialpublication800-64r2.pdf

3.   http://nvlpubs.nist.gov/nistpubs/SpecialPublications/NIST.SP.800-160.pdf

4.   http://csrc.nist.gov/publications/drafts/nistir-8179/nistir-8179-draft.pdf

5.   https://www.nist.gov/sites/default/files/documents/cyberframework/cybersecurity-framework-021214.pdf

6.   https://www.nist.gov/sites/default/files/documents/cyberframework/cybersecurity-framework-021214.pdf

Identify Stakeholder Assets & Authorization Boundary

During the preparation stage, the ISSO should ask what all the system components are. The ISSO is the eyes of the organization.

The organization must identify the assets that need protection. This includes information about the mission/business that the system supports. The organization needs to know what happens if the system is unavailable and how the system interacts with other systems. Most of this information will already exist from the ISSO's perspective unless it is a new system. For new systems, this kind of information will be in the concept of operations, technical system data about the mission the system supports, diagrams, hardware, software components lists, proposed site data, and other supporting documents. The more the ISSO has, the better. But if the ISSO has none of this, they should know that this is the organization's task, not theirs to complete alone. This is information that the organization has to figure out as a team. The ISSO can help but cannot take this on by themselves.

To fully identify the assets of a system, the organization needs to identify the authorization boundary. The authorization boundary defines the official systems and connections where the assets end. If the organization does not clearly define the authorized boundary, the asset could include other components that don't belong to it. An ISSO can see the boundary by the assets network diagrams, software/hardware list, and reviewing the concept of operation documents. Although this is the responsibility of the CIO, System Owners, and other high-level roles, it usually falls on technical experts and ISSOs to document the assets and their authorization boundary. The best thing you can do is have high-level network diagrams and confirm the boundary with a subject matter expert (i.e., network engineer, system admin, enterprise architect).

***References:** NIST Special Publications800-39[1] (Organization Level),800-64[2],800-160[3] (Stakeholder Needs and Requirements Definition Process); NIST Interagency Report8179[4] (Criticality Analysis Process C);Cybersecurity[5]Framework[6] (Core Identify Function);NARA[7]CUI[8]Registry[9].*

Identify the Types of Information to be Processed, Stored, and Transmitted by the System

One important step that is often skipped is identifying the information types of the system. What is the classification of the data? Is it secret, proprietary, PII, health care records, or satellite data? What data is being processed, stored, and transmitted?

The organization must identify the types of information needed to support the mission and business functions. You will need this information to categorize the system (if the system is not already categorized). This is all the data the ISSO has to get from the organization. The ISSO can help identify information types by making suggestions, but ultimately, the organization has to decide how the data is identified. This process is vital to developing security and privacy plans for the system and a precondition for determining the security

---

1. *http://nvlpubs.nist.gov/nistpubs/Legacy/SP/nistspecialpublication800-39.pdf*

2. *http://nvlpubs.nist.gov/nistpubs/Legacy/SP/nistspecialpublication800-64r2.pdf*

3. *http://nvlpubs.nist.gov/nistpubs/SpecialPublications/NIST.SP.800-160.pdf*

4. *http://csrc.nist.gov/publications/drafts/nistir-8179/nistir-8179-draft.pdf*

5. *https://www.nist.gov/sites/default/files/documents/cyberframework/cybersecurity-framework-021214.pdf*

6. *https://www.nist.gov/sites/default/files/documents/cyberframework/cybersecurity-framework-021214.pdf*

7. *https://www.archives.gov/cui*

8. *https://www.archives.gov/cui*

9. *https://www.archives.gov/cui*

categorization. NIST 800-60 v2 has the best guidance for this information.

References: *OMB A-130; NARA CUI; SP 800-39 (System Level); SP 800-60 v1; SP 800-60 v2; NIST CSF (Core Identify Function).*

Register the System with Organizational Program or Management Offices

Organizations typically have a list, a system, and a database where assets are listed. This is sometimes called "system registration." This list of assets is reported to a higher organization. This is especially important for newly created systems.

The ISSO should know the status of new and updated system registration. Does the organization need to upload data to a systems registration? Some organizations do not. But if they do, the ISSO can check to see if the system has already been loaded there and make sure the data is still up to date.

Categorize

"Categorize" refers to security and privacy categorization to rank how important the system is to the organization. It is the organization that makes the decision and final approval on the security categorization of the system.

In the Categorize step, the organization describes the system and decides the level of impact the system has. This categorization will determine what security and privacy control the system will have. This process determines the risk by looking at how vital confidentiality, integrity, and availability is.

Objectives of security categorization:

- Describe the characteristics of the system in the system security plan (SSP)

- Review and approve the security categorization results and decision

Describe the characteristics of the system

From the ISSOs perspective, the system characteristics are put into a system security plan (aka an SSP). The ISSO should work closely with the subject matter experts (SMEs) to get an accurate picture of the system. The SSP usually has an executive summary to give a short, concise description of the function and form of the system(s).

Within the SSP, the ISSO will need more technical details necessary to satisfy the security controls. These details include (but are not limited to) privacy plans/policies, network diagrams, ports, protocols and services, what systems it is directly connected with, software, hardware lists, stakeholders, network and user rules, and many other details.

References: *NIST Special Publication800-18[1];Cybersecurity[2]Framework[3] (Core Identify Function).*

1. *http://nvlpubs.nist.gov/nistpubs/Legacy/SP/nistspecialpublication800-18r1.pdf*

2. *https://www.nist.gov/sites/default/files/documents/cyberframework/cybersecurity-framework-021214.pdf*

3. *https://www.nist.gov/sites/default/files/documents/cyberframework/cybersecurity-framework-021214.pdf*

Review and Approve of the Security Categorization Results and decision

Unless the system is very new, the security categorization of a system is usually already done by the organization. But the ISSO needs to understand how Categorization is done (See FIPS Appendix below for details of security categorization).

If the ISSO has gathered all the information they need in the "Preparation" step If the ISSO has gathered the information they need in the "Preparation" step, security categorization will be known. If it is not, they will have everything they need to help the organization determine the information types, the classification, what systems are in the authorization boundary, and who to call to get more information.

For information systems that process personally identifiable information (PII) privacy data, the senior agency official for privacy (sometimes called a "privacy manager") reviews and approves the security categorization results and decision before the authorizing official's review.

The authorizing official (AO) or a designated representative ensures that the security category selected for the information system is consistent with the mission and business functions and will have adequate protection. The AO reviews the categorization results and decisions from an organization-wide perspective, including how the decision aligns with the categorization decisions for all other organizational systems. The AO collaborates with the CIO and risk management/risk executive (function) to ensure that the categorization decision for the system is consistent with the organizational risk management strategy and satisfies requirements for high-value assets.

The final decision of the system security categorization is put into the system registration (if any).

References: FIPS 199; SP 800-30; SP 800-39 (Organization Level); SP 800-160 v1 (Stakeholder Needs and Requirements Definition Process); CNSSI 1253; NIST CSF (Core Identify Function).

How does the organization Categorize the System?

All the instructions an organization needs to determine the security categorization of a system and its information are detailed in *FIPS 199, Standards for Security Categorization of Federal Information and Information Systems,* and *NIST SP 800-60 Volume 2, Guide for Mapping Types of Information and Information Systems to Security Categories.* FIPS 199 describes how systems are categorized, and NIST 800-60 describes types of information and suggests security categorization for each. Examples of types of information include Public, Space Exploration, Inspection & Audit, Federal grants, Military operations, Criminal Incarceration, Advising, and consulting (for more, see NIST 800-60 volume 2).

The organization may develop its process for determining the security categorization. Still, it will always come down to determining what type of information is on the system and the security objectives of the information. That means they need to consider the confidentiality, integrity, and availability of the information in question. They need to determine the impact on the organization if the information is leaked, corrupt or unavailable.

The importance of confidentiality, integrity, and availability can be low, moderate, or high. For example, a web server might have a low impact level for confidentiality because the organization's main business or mission is not directly affected by data on the webserver getting leaked. May be the availability of the webserver is more important, so that is considered moderate impact level for the availability.

"System A" is a web server with information assigned a low or perhaps NO confidentiality (the information is already public). The public information needs a moderate level of integrity because they want the public information published on the site to be accurate. The organization also wants the public information to have at least a moderate level of availability. Since System A has an impact level, it now becomes a system with a "moderate" security categorization.

SC (security categorization) public information = {(confidentiality, NA), (integrity, MODERATE), (availability, MODERATE)}

"Server B" is a protected database that collects real-time information on troop movements in warzones. If the military logistics information on this system were leaked to the opposition, this would seriously impact the organization and personnel. Troops could die. The information is classified. The confidentiality is HIGH. The organization is ok if users with access to the information cannot access it for a few days, so availability is considered a low level of impact.

> SC (security categorization) military logistics information = {(confidentiality, HIGH), (integrity, MODERATE), (availability, LOW)}

The way the organization determines which systems have a certain security categorization and impact level is based on the FIPS 199 and NIST 800-60; the process of security categorization is done in three steps:

Step 1: What are the information types on the system?

Organizations can use the NIST 800-60 volume 2 to understand what security categories they suggest for each information types. The system can have several information types on it.

Step 2: What is the security categorization of each information type?

Once the organization determines the information types, they need to figure out the security categorization of each information type. This is done by figuring out the impact of the security objectives on the information.

> Information Security Category = {(confidentiality, LOW), (integrity, LOW), (availability, LOW)}

In this equation, the information security category is equal to the highest level of impact among security objectives for information types on the system. Taking the highest impact level among information types on a system is a "high watermark." Knowing whether the high watermark is LOW, MODERATE, or HIGH will determine which security controls to select as a baseline and which security enhancements are necessary. Security enhancements and baselines are in the NIST 800-53.

note: This is different for national security systems (NSS). They do not use the high watermark method. NSS systems use an overlay to determine what level to give each control

Step 3: Determine the overall security categorization of the system based on the information types?

Information system Security Category = {(confidentiality, impact), (integrity, impact), (availability, impact)}, where the acceptable values for potential impact are low, moderate, or high. Take the highest impact level for the security category of the system.

Select

Once the security categorization has been determined, the organization can select the appropriate security controls.

The purpose of the *Selection* step is to identify, select, tailor, and document the security and privacy controls necessary to protect the system and the organization at a level equal to the risk it will have on the organizational operations and assets, individuals, other organizations, the industry or the nation.

Objectives of selecting security controls for a system:

- Select the security and privacy controls necessary to protect the system equal to the risk it will have on the organization

- Tailor the controls selected for the system and the environment of operation

- Allocate security and privacy controls to the system and the environment of operation

- Document the controls for the system and environment of operation in security and privacy plans

- Develop and implement a system-level strategy for monitoring control effectiveness that is consistent with and supplements the organizational continuous monitoring strategy

- Review and approve the security and privacy plans for the system and the environment of operation

Select the Security and Privacy Controls

As an ISSO, when you come to an already established system, the system has been categorized, and it has a set of security controls. For these systems, the ISSO is working on making sure those security controls are in place, implemented, and documented.

If it is a new system, the ISSO will be a part of the organization's effort to select security and privacy controls. This is based on the Categorization of the system. The ISSO will look to the NIST SP 800-53 for the baseline security controls that need to be addressed and the organization's security policy for additional controls. This is not done in a vacuum, with the ISSO deciding what is selected. The organization does this with the ISSO as part of that organization.

The organization can use a baseline control selection approach or an organization-generated control selection approach.

> **Baseline control selection approach:** This method uses control baselines, which are pre-defined sets of controls specifically assembled to address the protection needs of a group, organization, or community of interest. These controls can be found in NIST SP 800-53. The selection of the controls is based on the organization's security categorization, needs, laws, executive orders, and policies.

> **Organization-generated control selection:** The organization uses its selection process to select controls without using the pre-defined set of security controls in the 800-53. This happens with organizations that have a specialized mission or business needs. For example, an advanced system of medical devices or a space-based weapon system. It might not be practical to use the pre-defined set of controls in these cases. National Security Systems (NSS) sometimes use this approach with security control overlays (CNSSI 1253).

In both cases, the organization uses some of the 800-53 controls, but they have to tailor the selection process and the controls themselves to fit the system's security needs. The controls are documented in a system security plan.

References: *FIPS* *Publications199[1],200[2];* *NIST* *Special Publications800-18[3],800-30[4],800-53[5],800-160[6] (System Requirements Definition,*

---

1.   *http://nvlpubs.nist.gov/nistpubs/FIPS/NIST.FIPS.199.pdf*

2.   *http://nvlpubs.nist.gov/nistpubs/FIPS/NIST.FIPS.200.pdf*

3.   *http://nvlpubs.nist.gov/nistpubs/Legacy/SP/nistspecialpublication800-18r1.pdf*

Architecture Definition, and Design Definition Processes), 800-161[7] (Respond and Chapter 3); NIST Interagency Report8179[8]; CNSS Instruction 1253; Cybersecurity[9] Framework[10] (Core Identify, Protect, Detect, Respond, Recover Functions; Profiles).

4.   *http://nvlpubs.nist.gov/nistpubs/Legacy/SP/nistspecialpublication800-30r1.pdf*

5.   *http://nvlpubs.nist.gov/nistpubs/SpecialPublications/NIST.SP.800-53r4.pdf*

6.   *http://nvlpubs.nist.gov/nistpubs/SpecialPublications/NIST.SP.800-160.pdf*

7.   *http://doi.org/10.6028/NIST.SP.800-161*

8.   *http://csrc.nist.gov/publications/drafts/nistir-8179/nistir-8179-draft.pdf*

9.   *https://www.nist.gov/sites/default/files/documents/cyberframework/cybersecurity-framework-021214.pdf*

10.   *https://www.nist.gov/sites/default/files/documents/cyberframework/cybersecurity-framework-021214.pdf*

Tailor the Controls Selected for the System

Security control baselines are not one size fits all. Instead, baselines are tailored to fit each environment. Each environment has different threats and different levels of risk. For example, a forensics lab dealing with sensitive personal data and evidence will need more protection of stored evidence integrity than an agency with a system that streams entertainment data to employees within their organization. The Amber alert systems that broadcast missing person data probably need to pay close attention to data integrity than a weather notification system for privately owned servers.

If one organization uses wireless, they will need to pay close attention to controls that address this. But an organization that does not use wireless can tailor these controls because they don't need them.

The organization can base tailoring on the following:

- Mission / Business functions
- Threats to the system
- Assets
- The site and environment

References: *FIPS 199; FIPS 200; SP 800-30; SP 800-53; SP 800-53B; SP 800-160 v1 (System Requirements Definition, Architecture Definition, and Design Definition Processes); SP 800-161 and Chapter 3; IR 8179; CNSSI 1253; NIST CSF (Core Identify, Protect, Detect, Respond, Recover Functions; Profiles).*

Allocate security and privacy controls to the system

Understanding security and privacy controls is essential to determine where the controls go and who manages them. The NIST 800-37 describes three types of controls:

- **System-specific controls**—controls that provide a security capability for a particular information system only (example: multifactor authentication on a Windows or Red Hat system)

   ○ System-specific controls only provide a security capability for a particular information system and are the primary responsibility of information system owners and their respective authorizing officials. An example of security control typically implemented as a system-specific control is IA-6, Authenticator Feedback, where the information system is designed to obscure the feedback of authentication information during the authentication process. Displaying asterisks when a user types in a password is an example of obscuring feedback of authentication information.

- **Common controls**—controls that provide a security capability for multiple information systems (example: Physical controls of all systems in a facility)

   ○ Common controls are security controls that can support multiple information systems efficiently and effectively as a common capability. When these controls are used to support a specific information system, they are referenced by that specific system as inherited controls. Many of the security controls needed to protect organizational information systems (e.g., contingency planning controls, incident response controls, security training and awareness controls, personnel security controls, physical and environmental protection controls, and intrusion detection controls) are excellent candidates for common control status. Information security program management controls may also be deemed common controls by the organization since the controls are employed at the organizational level and typically serve multiple information systems.

● **Hybrid controls**—controls that have both system-specific and common characteristics (example: personnel security - ALL personnel entering the facility must have a badge)

○ Hybrid controls are security controls where one part of the control is deemed common, and another is system-specific. The organization may choose, for example, to implement the Security Awareness security control (AT-2) as a hybrid control with general organization-wide security awareness training provided as a common capability with focused security awareness training provided for the specific information system.

The organization assigns controls as system-specific, hybrid, or common and allocates the controls to the system elements (i.e., machine, physical, or human elements). Each element provides the controls.

References: *SP 800-39 (Organization, Mission/Business Process, and System Levels); SP 800-64; SP 800-160 v1 (System Requirements Definition, Architecture Definition, and Design Definition Processes); NIST CSF (Core Identify Function; Profiles); OMB FEA.*

Document the controls for the system in security and privacy plans

One of the main jobs of an ISSO in the RMF process is to document the security controls for the system(s) they are assigned to. This is done in a system security plan and security policy. A great reference for a system security plan is *NIST 800-18, Guide for Developing Security Plans for Federal Information Systems.* 800-18 provides the format of the system security plan. Security plans contain an overview of the security and privacy requirements for the system and the controls selected to satisfy the requirements. The plans describe the status of each control on the system and how it is applied. This description can include technical, operational, or executive summaries of how the control is applied.

Documentation allows the organization to have continuity and allows the ability to trace the decisions made for the system.

References: FIPS 199; FIPS 200; SP 800-18; SP 800-30; SP 800-53; SP 800-64; SP 800-160 v1 (System Requirements Definition, Architecture Definition, and Design Definition Processes); SP 800-161 (Respond and Chapter 3); IR 8179; CNSSI 1253; NIST CSF (Core Identify, Protect, Detect, Respond, Recover Functions; Profiles).

Develop a monitoring strategy

The ISSO should work with the subject matter experts, information system owners, and other stakeholders to develop a plan and deadline to implement and monitor controls. It's the organization that comes with the frequency of continuous monitoring.

An effective continuous monitoring strategy will streamline the system development life cycle (SDLC). It will follow the system's design, implementation, ongoing maintenance, and decommission.

References: *SP 800-30; SP 800-39 (Organization, Mission or Business Process, System Levels); SP 800- 53; SP 800-53A; SP 800-137; SP 800-161; IR 8011 v1; CNSSI 1253; NIST CSF (Core Detect Function).*

Review security and privacy plans for the system

As the team develops documentation, the ISSO should ensure all relevant stakeholders know what is going on. There should be a constant review of the security and privacy plans. This is why many organizations put the system security plan into a database or secure, shared website to make it available to all constantly.

The documentation of the selected security controls is written by a cybersecurity role such as the ISSO and often reviewed and approved by other departments within the organization, such as the office of the CIO or a risk management team, or maybe an information system security management office.

References: *SP 800-30; SP 800-53; SP 800-160 v1 (System Requirements Definition, Architecture Definition, and Design Definition Processes).*

How does the organization know what security controls to Select?

The best guidance on selecting security controls is in *FIPS 200, Minimum Security Requirements for Federal Information and Information Systems,* and *NIST SP 800-53, Security and Privacy Controls for Information Systems and Organizations.* One of the best resources is the NIST site: csrc.nist.gov, which breaks down each control into Low, Moderate, and High. There are three steps to selecting a security control:

Step 1: Use the system security categorization to determine the baseline security controls (Reference: NIST SP 800-53:

Low:

The unauthorized disclosure of information, modification, or disruption of access could have a limited adverse effect on organizational operations, assets, or individuals.

Moderate:

The unauthorized disclosure of information, modification, or disruption of access could severely adversely affect organizational operations, assets, or individuals.

High:

The unauthorized disclosure of information, modification, or disruption of access could have a severe or catastrophic adverse effect on organizational operations, assets, or individuals.

Step 2: Add or Remove Security and Privacy controls to fit the system

Not all controls need to be used on a system. NIST 800 baseline controls are not meant to be one size fits all. For example, if none of the sites use wireless technology, there will be no need to use AC-18, wireless access. The organization might have special healthcare information that requires

HIPAA. They may need to create controls that cover protected healthcare information.

Step 3: Document the Controls in a System Security Plan

The selection of controls is not made by word of mouth; it must be documented. Documentation of security controls is done in forms such as a system security plan.

Implement

Once the organization selects the controls, it is time to implement the security features on the systems, sites, and organization's behaviors.

Objectives of implementing security controls:

- Implement Controls
- Update Control implementation information

Implement Controls

Implementation of security controls is not just technical. For example, it's not just configuring systems to have 14-character passwords. Implementation is also making sure the facilities windows on the first floor are locked up at night. It also means ensuring the organization has signed off on a policy that addresses employees who bring their hard drive to work. Understanding the types of security controls can help an ISSO understand what needs to be done.

Security control types include system-level, hybrid, and common controls. These types define who manages each security control and the reach of the controls (see Select for more and system-level, hybrid, and common controls). We could also categorize security controls by how they work. The NIST documents sometimes mention: technical, operational, and management controls.

Technical controls:

Technical controls are what most people think of when we talk about implementation. These controls deal with installing, configuring, and enabling security and privacy features on servers, laptops, mobile devices, and other information systems. The technical controls are implemented by system administrators, system engineers, field technicians, or other purely technical roles. But it can also be given to the ISSO role. It depends on the organization and how they define its roles.

Regardless of who implements the technical controls, they will all use the same resources to help figure out how to do it. The main source is the vendors. This means Microsoft, Oracle, SUSE, Red Hat, Cisco, or whatever organization created the system you are using. Other resources include:

- Security Technical Implementation Guides - https[1]://[2]public[3].[4]cyber[5].[6]mil[7]/[8]stigs[9]/[10]

---

1. https://public.cyber.mil/stigs/

2. https://public.cyber.mil/stigs/

3. https://public.cyber.mil/stigs/

4. https://public.cyber.mil/stigs/

5. https://public.cyber.mil/stigs/

6. https://public.cyber.mil/stigs/

7. https://public.cyber.mil/stigs/

8. https://public.cyber.mil/stigs/

9. https://public.cyber.mil/stigs/

10. https://public.cyber.mil/stigs/

● NSA Security Guides - nsa.gov/What-We-Do/Cybersecurity/Advisories-Technical-Guidance/

● NIST National Checklist Program - https[11]://[12]ncp[13].[14]nist[15].[16]gov[17]/[18]repository[19]

There are a few things that you should always do when implementing the technical controls:

● Make sure you have a subject matter expert to lead the work that needs to be done.

● Always back up the system before you start

● Have a system restore ready to go before you accidentally lock yourself out of the system

● Do NOT implement all the controls at once and cross your figures, hoping it will work after you reboot

● RTMF

● Get written permission before you implement controls

● Examine the old implementation if any

Operation controls:
Operational controls include ensuring security training is performed regularly, ensuring all employees wear badges in certain areas, and ensuring personnel knows and follows policies.
Management controls:

---

11. https://ncp.nist.gov/repository

12. https://ncp.nist.gov/repository

13. https://ncp.nist.gov/repository

14. https://ncp.nist.gov/repository

15. https://ncp.nist.gov/repository

16. https://ncp.nist.gov/repository

17. https://ncp.nist.gov/repository

18. https://ncp.nist.gov/repository

19. https://ncp.nist.gov/repository

Some controls focus on how the organization manages security and privacy. These controls ensure the organization has policies approved by upper management and clear procedures so that personnel have guidance. You will notice that every family of controls starts with management controls dealing with policy and procedures.

No matter what controls you are tasked to do, make sure you give yourself extra time just in case the implementation does not go well. It usually does not go right the first time.

There are many places to get information on how to implement the controls. STIGS, NSA, and Vendor sites are the main sources for implementing the security controls.

Update Implement Control Information

Once controls are in place, the documentation will need to be updated. Some security features will not be able to be implemented as originally expected. Some controls will not be able to be implemented at all. This will require an update.

As an ISSO, this is often the main job. Documenting the controls is a big job that organizations neglect. Nobody realizes how important it is until there is a significant audit or a lack of continuity between one group of personnel and another. For this reason, there is often politics, blame, and stress associated with updating the implementation of controls. Everyone wants to point fingers, but this is something the organization is responsible for. The ISSO might upload the data into "eMASS" or some other flavor of the month database or spreadsheet. Still, they have to gather this information from different parts of the entire organization. For example, suppose the backup process has changed and moved from SystemBackUPs version 2 to SystemBackUPs version 4. In that case, the ISSO needs to talk to the backup administrators responsible for this system to get a summary of what has changed from Incremental backups to Full backups. In this case, the organization would have had to have a change management process designed to inform all the stakeholders, including the friendly neighborhood ISSO, so they can make any changes necessary once the project is complete.

Assess

For the assessment step, the organization looks at the controls to determine if they have been implemented effectively. Depending on the assessment, the assessor might look at all aspects of security surrounding the system. Inside out and outside in. This includes technical controls on applications, the individual systems hosting the applications, the network connections of the hosting systems, the internal network, and anything within the system boundary.

The organization should also consider operational controls such as documentation detailing the policy, process, and procedures—the security awareness of the personnel, preparation for security incidents and disasters.

The physical security of the system must also be observed. If the system is not Physically protected, all technical and operational controls can be bypassed.

Objectives of assessing the security controls of a system:

- Assessor Selection
- Assessment Plan
- Control Assessments
- Assessment Reports
- Remediation Actions
- Plan of Action and Milestones

Assessor Selection

The organization needs to select the appropriate assessor or assessment team for the type of control assessment. The type of team depends on the type of assessment. The organization has internal self-assessments; there may be industry-level audits, risk assessments based on a recent breach, or significant changes in an information system.

Assessment Plan

The assessor coordinates with the organization to develop, review, and approve plans to assess implemented controls. The assessor will detail the organization's requirements on the security assessment plan (SAP). They will document who is doing the assessment, what kind of assessment they are doing when it will be done, what systems and controls will be covered, and why it is being conducted.

Control Assessments

The assessor evaluates the controls according to the assessment procedures described in SAP. Assessment methods consist of interviews, testing, and observations. The assessor can run network scans, do physical security checks, and review documentation to determine if the security controls are implemented.

Assessment Reports

As the assessor does the assessment, they have to document the process. They prepare the assessment reports by documenting the findings and recommendations from the controls assessed. They distribute this documentation to the organization.

Remediation Actions

Once the organization receives the assessment report, it can conduct initial remediation actions on the controls and reassess remediated controls. They usually review all the results and determine what can be fixed. Anything that cannot be fixed within the confines of the vulnerability management schedule will need to be documented in a risk response. A risk response may be risk acceptance, acknowledgment, or a plan of action and milestone (POAM).

Plan of Action and Milestone (POAM)

For all findings from the assessment that cannot be remediated, the organization will prepare a plan of action and milestones. A POAM breaks down what needs to be remediated, what security controls are affected, who will fix them, and when. A POAM may also include compensating controls; security features cover the findings found to lower the risk to the organization. The plan to fix the security controls must have milestones to get done on time with the organization's resources.

Authorize

It feels like everything comes down to the authorization step. Everything you have done up to this point hinges on the authorization of the system. During the authorization step, the heads of the agency (CEO, CIO, commander, authorization official, etc.) consider all the data from the assessment to determine if they will accept the risk in writing. All the security controls have to be applied correctly, security appropriately categorized, and risk-managed to an acceptable level.

Objectives of authorizing an information system:

- Authorization Package
- Risk Analysis and Determination
- Risk Response
- Authorization Decision
- Authorization Reporting

Authorization Package

After assessing the system, you will have everything you need to go forward with an authorization. You will need to assemble the authorization package and submit it to the authorizing official for a decision.

The authorization package consists of (but is not limited to):

- System Security Plan
- System Assessment Plan
- Security Assessment Report
- Plan of Action and Milestone
- Risk Acceptance
- Artifacts: supporting evidence

Risk Analysis and Determination

The organization will determine the risk of the system by looking at how the controls have been implemented. The focus will be on things that stand out: high and critical controls that could not be applied, exceptions to the policy, plan of action and milestones, risk acceptance documents, and residual risks. An example of a residual risk would be a system that could not implement the multifactor authentication required by the organization, so they implemented a username password. Even though the username and password are secure, there may be a low level of risk associated with how the username and password are managed. The risk is that multifactor authentication is way more secure than username and passwords, and perhaps the implementation of the username and password is only 12 characters long. It is supposed to be 14 characters at least. The organization now has to accept that leftover (residual) risk.

After analyzing all the security controls, the organization will determine the risk. They will decide how to respond to the risk.

Risk Response

The organization will identify and implement a preferred course of action in response to the risk determined. In over words, they will decide to do with each find left on the system. There are several risk responses they can use:

- Avoiding – eliminate the cause of the risk
- Mitigating – reducing risk with other security controls methods
- Risk Accepting – documenting and acknowledging the risk and finding compensating controls for the remaining risk when possible.
- Risk Transferring – delegating or moving the risk to another organization
- Risk Sharing – sharing the risk with another organization

Authorization Decision

The full acceptance of risk cannot be delegated to other officials. The authorizing official. They use all the data to make a decision. And once the decision is made, they formally document the decision with an authorization letter or some document that indicates the risk has been accepted.

Authorization Reporting

Organizations usually have some way of reporting the authorization of a system. This is so that risk decisions can be viewed in the context of other systems in the organization or among other organizations.

Monitor

The process doesn't stop with the authorized step. Once the system is authorized, it must be continuously monitored. The organization regularly checks the documentation, technical, management, operational, and physical security controls. Privacy controls are constantly checked. This ensures the organization maintains a certain level of security and privacy posture.

Objectives of continuous monitoring:

- System and Environment Changes
- Ongoing Assessments
- Ongoing Risk Response
- Authorization Package Updates
- Security and Privacy Reporting
- Ongoing Authorization
- System Disposal

System and Environment Changes

Monitor the information system and its environment of operation for changes that impact the security and privacy posture of the system. Systems and environments of operation are in a constant state of change. These changes can happen in the technology, human resources, the physical location, and many other factors. Changes to the technology are upgrades to hardware, software, or firmware. The human resources can also be modified by staff turnover or a drastic reduction in force. The systems can change location or be affected by natural disasters. These things affect the physical and environmental controls.

Ongoing Assessments

Assessments will be a part of the organization's continuous monitoring strategy. After authorization, the organization will assess all controls regularly to test if they have been appropriately implemented (or at all). The organization determines the frequency.

Ongoing Risk Response

The organization will monitor ongoing risk responses and maintain a plan of action and milestones created for controls. In some cases, they will need to create new risk responses and POAMs. According to the NIST 800-37, the authorizing official determines the appropriate risk response to the assessment findings or approves responses proposed by the system owner and common control provider.

Authorization Package Updates

The artifacts within the authorization package need to be updated continuously. For example, as systems change, documents like the system security plan will need to change too. The plan of action and milestones are updated, removed, and renewed.

Security and Privacy Reporting

Risk assessments determine the system's security and privacy posture and report this to the authorizing official and other organizational officials. This is done on an ongoing basis following the organizational continuous monitoring strategy. This can be done by a third-party organization or internally.

Ongoing Authorization

The organization must ensure that they maintain the security posture that was approved during the system's authorization. They will do this by frequent reviews of the system's security and privacy posture. They need to determine whether the risk of the system is still acceptable. Regular network scans, policy/procedure reviews, and risk assessments are done at regular intervals mandated by the organization to monitor the authorization continuously.

System Disposal

There will be obsolete systems and end of life with ongoing security posture and authorization reviews. This may start the process of the disposal of the system. The organization will need to implement a system disposal strategy and execute necessary actions when a system is removed from operation. This might include media sanitization, configuration management and control, component authenticity, and record retention.

RMF ROLES:

Authorizing Official (AO) - (NEEDED) Responsible for the entire system. Gives final approval;

Takes risks. Either above the CIO, the CIO or appointed by the CIO

(aka - DAA, Designated Approval Authority, Commander, CIO, AO). **From the ISSO's perspective:** This is arguably the most important person who will need to be a part of the process. That AO approves, disapproves, or takes the risk of the system going through the RMF process.

Chief Acquisition Officer (Acquisitions) - (NOT ALWAYS NECESSARY) Monitors the organization's performance of acquisitions. Usually, part of the CIO office or entire acquisition team; is necessary for buying lots of new equipment. **From the ISSO's perspective:** This position is only relevant if there is a lot of new equipment to purchase on the license to renew. The name "chief acquisition officer" will probably not be used, but it will be some individual or group who performs this work.

Chief Information Officer (CIO) - (SOMETIMES NEEDED) Very important executive responsible for information technology issues, including

security, acquisitions, sustainability, and IT risks (aka - Authorizing Official, DAA, usually

an entire office of people). **From the ISSO's perspective:** The CIO is very important because security falls under the CIO. The CIO is not directly involved in the RMF process in some instances. Each organization handles the AO role differently, but usually, the CIO is involved with the RMF process. They may not be in any meetings, but you will deal with a representative from the CIO's team at some point.

Common Control Provider - (NEEDED) individual or group responsible for implementing, assessing, and monitoring common controls. Common Controls are controls that most systems within the organizations have to have (i.e., physical controls, environmental controls, training). **The ISSO's perspective:** The Common Control Provider is usually explained in the security control policy. The security control policy should explain physical security, personnel security, access controls, and other things that apply to ALL people and systems in the organization. If there is NO security control policy, the ISSO and ISSM may need to get with the information system owner and customer and create one. That document will cover common controls.

Control Assessor - (NEEDED) Individual or group that conducts a comprehensive assessment of the security controls. Need some level of independence from the authorizing official. Sometimes it is a unit working for the AO (risk management team, assessment team, risk assessment team, auditors, audit team). **The ISSO's perspective:** The Control Assessors are scary people because they report problems directly to the CIO/AO. But if you do what you're supposed to do, they don't have to cause any stress. They can be your best asset. If you ask them what they need from you before they even get there, they are usually willing to give you a checklist and forms to fill out, and sometimes they will even glance at your system before they start officially and tell you they want to fix.

ENTERPRISE ARCHITECT - (SOMETIMES NEEDED) These are subject matter expert on the system's layout. They understand how the system connects to the larger enterprise. From the ISSO's perspective: An Enterprise Architect is only necessary for assistance with integration, managing sizeable interconnected systems, or developing large systems.

Head of Agency - (NEEDED if they are the AO) Responsible for information security protections that protect the overall organization in the interests of the organization. (aka - DAA, Designated Approval Authority, Commander, AO). **The ISSO's perspective:** The Head of Agency is so high up that you don't even have meetings with them in large or even medium-sized corporations. Even if they are the AO, they will have an AO office that will deal with you.

Information Owner Or Steward - (NEEDED) This is an organizational official with statutory, management, or operational authority for specified information. They are responsible for ensuring that information follows policy and procedures when generated, collected, processed, stored, and disposed of (aka information system officer, ISO, data owner, system owner). This is very similar to a **Mission/Business Owner** whose focus is a specific line of business (contract supporting a system). **From the ISSO's perspective:** The information owner will need to be in on every process level. They are like representatives of the system. They speak to the office of the AO, the users of the system, and you (the ISSO). They are one of the most influential people you need to talk to because they know what AO, Assessors, and users want to see. These people are not always called Information Owners. But you will know them because they are not entirely in charge but in the middle of everything involving that system. Listen to them.

System Security / Privacy Officers - (NEEDED) An organization should have a system/privacy officer who ensures each system fits with the organization's operation security/privacy posture. **From the ISSO's perspective:** The ISSO is part of this role, usually for one or more systems within an organization. They might work with other system security/privacy officers. These people help start, maintain and complete the RMF process.

Security / Privacy Architect - (SOMETIMES NEEDED) These are individuals, groups, or organizations that ensure the security and privacy requirements are being met for all systems added to the Enterprise architecture. **From the ISSO's perspective, these people might be needed if the system going through the RMF process is** integrated into an existing enterprise. There may be additional security/privacy documents or processes necessary. Interaction with them might be as simple as an email or phone call.

Senior Accountable Official for Risk Management - (NEEDED) It is a fancy title for the person (or group) in your organization that leads the direction of the risk management process. In most cases, this is appointed by the head of the organization and under the office of the CIO. They lead the Risk Management (function). It is called a risk management "function" because it is not necessarily a single group or person, but it is something that must be done regardless of group, title, or role. **From the ISSO's perspective:** This could play a significant role in that all processes and documentation might go through this person, their representatives, or their office. They will look through your documentation, evidence, and process to determine if you are in line with the rest of the organization before sending the documents forwarded to the AO. An ISSO wants to make sure they are completely in line with the office that is in charge of the risk management process.

Senior Agency Information Security Officer - (NEEDED SOMETIMES) They are an organizational official responsible for carrying out the chief information officer security responsibilities under FISMA and serving as the primary liaison for the chief information officer to the organization's authorizing officials, system owners, common control providers, and system security officers. They usually have several hats, including but not limited to: risk executive function, head of common control providers, and system owners. **From the ISSO's perspective:** This position is defined by what they do. It will usually be high-level security that does EVERYTHING for the CIO. They are the main people you talk to for privacy, common controls, security control assessments, risk executive functions... and anything else you can think of.

Senior agency official for privacy - (NEEDED SOMETIMES) This senior official or executive with agency-wide responsibility and accountability for ensuring compliance with applicable privacy requirements and managing privacy risk. **From the ISSO's perspective:** You might need to coordinate with this person or their office if you are working on a system requiring a privacy assessment or if you have questions about the organization's privacy policies.

System Administrator - (NEEDED) The System Administrator is an individual, group, or organization responsible for setting up and maintaining a system or specific components of a system. System administrator responsibilities include, for example, installing, configuring, and updating hardware and software; establishing and managing user accounts. This is the person who touches the system. **From the ISSO's perspective:** Nothing on the system will get done without the system administrator. You need to contact this person or organization from day one, even just to let them know what will happen. In some cases, the ISSO is also doing some system administrator work.

System Owner - (NEEDED) A system owner is an organizational official responsible for the procurement, development, integration, modification, operation, maintenance, and disposal of an organizational system. The system owner is responsible for addressing the operational interests of the user community (i.e., users who require access to the system to satisfy mission, business, or operational requirements) and for ensuring compliance with security requirements. **From the ISSO's perspective:** Some organizations interchangeably use the information system owner, information owner, and system owner. The system owner makes sure that the user experience is not interrupted by the security. They also want to ensure that there is adequate security in place. The system owner may conduct an operational test to ensure that the system works appropriately after all security controls are applied.

System User - The system user is an individual or (system) process acting on behalf of an individual authorized to access organizational information and systems to perform assigned duties. **From the ISSO's perspective:** A user representative, system owner, or information owner may participate in the RMF process on behalf of the user. They want to ensure that the system is still operational after implementing the security controls. They also want to ensure that user data is protected.

References: *NIST*   *Special Publication 800-37; NICE[1] Cybersecurity[2] Workforce[3] Framework[4].*

---

1.   *https://www.nist.gov/itl/applied-cybersecurity/nice/resources/nice-cybersecurity-workforce-framework*

2.   *https://www.nist.gov/itl/applied-cybersecurity/nice/resources/nice-cybersecurity-workforce-framework*

3.   *https://www.nist.gov/itl/applied-cybersecurity/nice/resources/nice-cybersecurity-workforce-framework*

4.   *https://www.nist.gov/itl/applied-cybersecurity/nice/resources/nice-cybersecurity-workforce-framework*

FIPS 199 - Standards for Security Categorization of Federal Information and Information Systems

Insecurity, we categorize systems to know how many security controls to put on them. For example, a web server with public information will have a different category than a classified surface-to-air mission system.

The security categories are based on the potential impact on an organization should certain events occur which jeopardize the information and information systems needed by the organization to accomplish its assigned mission, protect its assets, fulfill its legal responsibilities, maintain its day-to-day functions, and protect individuals. Security categories are to be used in conjunction with vulnerability and threat information in assessing the risk to an organization. FIPS 199

Security Objectives

The FISMA defines three security objectives for information and information systems:

CONFIDENTIALITY

"Preserving authorized restrictions on information access and disclosure, including means for protecting personal privacy and proprietary information..." 44 U.S.C., Sec. 3542. A loss of confidentiality is the unauthorized disclosure of information.

INTEGRITY

"Guarding against improper information modification or destruction, and ensuring information non-repudiation and authenticity..." 44 U.S.C., Sec. 3542. A loss of integrity is the unauthorized modification or destruction of information.

AVAILABILITY

"Ensuring timely and reliable access to and use of information..." 44 U.S.C., SEC. 3542. A loss of availability is the disruption of access to or use of information or an information system. Potential Impact on Organizations and Individuals.

(3) Levels of Potential Impact

FIPS Publication 199 defines three levels of the potential impact on organizations or individuals should there be a security breach (i.e., a loss of confidentiality, integrity, or availability). Applying these definitions must occur within the context of each organization and the overall national interest.

LOW

The potential impact is LOW if the loss of confidentiality, integrity, or availability could be expected to have a limited adverse effect on organizational operations, organizational assets, or individuals. A limited adverse effect means that, for example, the loss of confidentiality, integrity, or availability might: (i) cause degradation in mission capability to an extent and duration that the organization can perform its primary functions, but the effectiveness of the functions is noticeably reduced; (ii) result in minor damage to organizational assets; (iii) result in minor financial loss; or (iv) result in minor harm to individuals.

MODERATE

The potential impact is MODERATE if the loss of confidentiality, integrity, or availability could be expected to have a severe adverse effect on organizational operations, organizational assets, or individuals. A severe adverse effect means that, for example, the loss of confidentiality, integrity, or availability might: (i) cause significant degradation in mission capability to an extent and duration that the organization can perform its primary functions, but the effectiveness of the functions is significantly reduced; (ii) result in significant damage to organizational assets; (iii) result in significant financial loss; or (iv) result in significant harm to individuals that do not involve loss of life or severe life-threatening injuries.

HIGH

The potential impact is HIGH if the loss of confidentiality, integrity, or availability could be expected to have a severe or catastrophic adverse effect on organizational operations, organizational assets, or individuals. A severe or catastrophic adverse effect means that, for example, the loss of confidentiality, integrity, or availability might: (i) cause severe degradation in or loss of mission capability to an extent and duration that the organization is not able to perform one or more of its primary functions; (ii) result in major damage to organizational assets; (iii) result in major financial loss; or (iv) result in severe or catastrophic harm to individuals involving loss of life or serious life-threatening injuries.

Security Categorization Applied to Information Types

The security category of an information type can be associated with user information and system information3 and can apply to information in either electronic or non-electronic form. It can also be used as input in considering the appropriate security category of an information system (see description of security categories for information systems below). Establishing an appropriate security category of an information type essentially requires determining the potential impact of each security objective associated with the particular information type.

The generalized format for expressing the security category, SC, of an information type is:

SC information type = {(confidentiality, impact), (integrity, impact), (availability, impact)},

where the acceptable values for potential impact are LOW, MODERATE, HIGH, or NOT APPLICABLE.

EXAMPLE 1: An organization managing public information on its web server determines that there is no potential impact from a loss of confidentiality (i.e., confidentiality requirements are not applicable), a moderate potential impact from a loss of integrity, and a moderate potential impact from a loss of availability. The resulting security category, SC, of this information type is expressed as:

SC public information = {(confidentiality, NA), (integrity, MODERATE), (availability, MODERATE)}.

EXAMPLE 2: A law enforcement organization managing sensitive investigative information determines that the potential impact from a loss of confidentiality is high, the potential impact from a loss of integrity is moderate, and the potential impact from a loss of availability is moderate. The resulting security category, SC, of this information type is expressed as:

SC investigative information = {(confidentiality, HIGH), (integrity, MODERATE), (availability, MODERATE)}.

EXAMPLE 3: A financial organization managing routine administrative information (not privacy-related information) determines that the potential impact from a loss of confidentiality is low, the potential impact from a loss of integrity is low, and the potential impact from a loss of availability is low. The resulting security category, SC, of this information type is expressed as:

SC administrative information = {(confidentiality, LOW), (integrity, LOW), (availability, LOW)}.

Security Categorization Applied to Information Systems

Determining the security category of an information system requires slightly more analysis and must consider the security categories of all information types resident in the information system. For an information system, the potential impact values assigned to the respective security objectives (confidentiality, integrity, availability) shall be the highest (i.e., high watermark) from among those security categories determined for each type of information resident in the information system.

The generalized format for expressing the security category, SC, of an information system is:

SC information system = {(confidentiality, impact), (integrity, impact), (availability, impact)},

The acceptable values for potential impact are LOW, MODERATE, or HIGH. Note that the value of not applicable cannot be assigned to any security objective in establishing a security category for an information system. This is in recognition that there is a low minimum potential impact (i.e., low watermark) on the loss of confidentiality, integrity, and availability for an information system due to the fundamental requirement to protect the system-level processing functions and information critical to the operation of the information system.

EXAMPLE 4: An information system used for significant acquisitions in a contracting organization contains sensitive, pre-solicitation phase contract information and routine administrative information. The management within the contracting organization determines that: (i) for the sensitive contract information, the potential impact from a loss of confidentiality is moderate, the potential impact from a loss of integrity is moderate, and the potential impact from a loss of availability is low; and (ii) for the routine administrative information (non-privacy-related information), the potential impact from a loss of confidentiality is low, the potential impact from a loss of integrity is low, and the potential impact from a loss of availability is low. The resulting security categories, SC, of these information types are expressed as:

SC contract information = {(confidentiality, MODERATE), (integrity, MODERATE), (availability, LOW)},

And SC administrative information = {(confidentiality, LOW), (integrity, LOW), (availability, LOW)}.

The resulting security category of the information system is expressed as:

SC acquisition system = {(confidentiality, MODERATE), (integrity, MODERATE), (availability, LOW)},

representing the high-water mark or maximum potential impact values for each security objective from the information types resident in the acquisition system.

About the Author

Hello, I'm Bruce Brown. I started in physical security in 1996 with the United States Airforce. My Air Force Specialty Code (AFSC) (aka Military Occupation Specialty Code or enlisted rating) was 3P0X1. We conducted all forms of physical security. We were tasked with providing force protection duties, guarding weapons, airbases, and Air Force personnel from potential dangers. This Air Force infantry job was eventually merged with law enforcement. I was a cop. It remains the most complicated and most stressful job I have ever had. I changed jobs and went into 3C01, Computer Systems Operations. After a few months of training, I was put on the help desk and then network engineering. I was so obsessed with computers and the tech industry that I volunteered to do additional work to learn. I got a few certifications (CompTIA A+, Network+, CCNA, and MCP), and I got a degree in information technology. I did a lot of jobs in the Air Force, but the last one I did was certification & accreditation.

I did not like Certification & Accreditation (now known as Risk Management Framework). It felt like pure paperwork. I wanted to be hands-on. I wanted to put networks together. When I got out of the Air Force and became a contractor, I decided I wouldn't even put certification & accreditation on my resume. For a few years, I did pure technical work. I was a field technician for mission systems; I did networking and system administration. We were overworked and underpaid, so I eventually left. I needed a new job fast, so I decided to put all the certification & accreditation work back on my resume. As soon as I did, I got lots of job offers paying 65% - 80% more to do less work.

The job I took needed me to maintain or create a system security plan. They also wanted me to get a CISSP certification in 6 months. Certification and accreditation evolved into what is now known as Risk Management Framework. I would be placed in the role of an Information System Security Officer. It was more than paperwork. The process focused more on managing risk. It was a lot of coordination with everyone developing the system. I realized that there is way less competition in Risk Management Framework. I think it's because most people don't know about it, and those who do know about it don't want to do it. It's not fun, and no one talks about it. But it pays, and you will not be out of a job.